CW01337469

MENTORING THE MACHINES

SURVIVING THE DEEP IMPACT OF
AN ARTIFICIALLY INTELLIGENT TOMORROW

PART ONE:
ORIENTATION

JOHN VERVAEKE

SHAWN COYNE

Story Grid Publishing LLC
P.O. Box 1091
Sag Harbor, NY 11963

First Edition Published August 2023

Edited by Leslie Watts

Additional Editorial Contributions from Tim Grahl and Danielle Kiowski

First Story Grid Publishing Paperback Edition
August 2023

For Information about Special Discounts for Bulk Purchases,
Please visit www.storygridpublishing.com

ISBN: 978-1-64501-082-1
Ebook: 978-1-64501-083-8

DEDICATION

For Our Children

ORIENTATION

"If a machine is expected to be infallible,
it cannot also be intelligent."

— Alan Turing (1912–1954)

"There isn't anyone you couldn't love
once you've heard their story."

— Sister Mary Lou Kownacki (1941–2023),
Order of Saint Benedict

1

WHO IS THIS PROJECT FOR?

This project is dedicated to expanding the general folk understanding of Artificial Intelligence (AI) in four progressive parts. One reader at a time.

We wrote it for "you."

It was inspired by and grew out of John Vervaeke's YouTube video essay *AI: A Cognitive Scientist's Warning* and John's companion video discussions, *Christianity & AI* with Ryan Barton, *AI = The Beast?* with Jonathan Pageau, *Conversation with John Vervaeke—AI Edition* with Jordan Hall, *Is AI the New Atom Bomb?* with Chloe Valdary, *Will AI Become Sentient?* with Johannes Niederhauser and Sean McFadden, and *Artificial Intelligence, The End of Humanity? or A New Beginning* with Alex Levy.

Here are the parts.

Part One: Orientation

Sets the stage for the investigation by framing the rise of ever more powerful forms of Artificial Intelligence as increasingly problematic.

Part Two: Origins

Describes the history of Artificial Intelligence and its

present state and proposes a goal state loosely defined as coexistent, communicative, coordinated, and cooperative with human existence, in other words aligned with their creators.

Part Three: Thresholds

Explains the theoretic nature and function of Artificial Intelligence. The good news is that cognitive science is not surprised by AI's present behavior. Moreover, we have robust frameworks to explain where it sits on a developmental model. As our predictions have thus far proven prophetic, with foresight we propose we can meet the machines at their critical stages of development and empower them to grow intelligently, rationally, and wisely in communion with us.

Part Four: Alignment

Summarizes our throughline arguments into a coherent recommended disposition toward the development of Artificially Intelligent, rational, and wise beings.

The goal of this project is to enable someone of limited familiarity with cognitive science's convergent integration of six core disciplines:

1. Linguistics
2. Philosophy
3. Anthropology
4. Neuroscience
5. Psychology
6. Computer Science

Or story science's integration of six core genres:

1. Worldview
2. Morality
3. Status

4. Action
5. Love
6. Horror

to follow the reasonable flow of Part One's global orientation, Part Two's historical "what got us here won't get us where we wish to be in the future" arguments, and Part Three's explanations of cognitive science's top-down emanation and bottom-up emergence of life, mind, and culture from matter. By the time you reach Part Four's proposed conclusions, we intend for you to have an optimal grip on the promise and peril of the probable arrival of artificially intelligent beings.

We'll be using the Feynman Technique, named after physicist Richard Feynman, who advised the valuable exercise for the scientist to use folk language to explain the essence (the structure, function, and organization) of their scientific theory.

Feynman himself famously used the concept of a ratchet to describe the complexities of quantum mechanics. And scientists have used his ratchet metaphor to describe threshold moments when novel, emergent phenomena are realized in biology, psychology, and sociology too.

As Gregg Henriques proposes in his Unified Theory of Knowledge model, the universe has ratcheted itself from

1. Matter to
2. Life to
3. Mind to
4. Culture

We've been teetering at the apex of that ratcheting for at least two hundred thousand years. AI may become a new form of matter into life into mind and culminate in culture, but the ratchet remains the same. It's a variant of a theme—a tale as old as time.

While we've taken great care generating this project, we will undoubtedly fail to meet all readers' expectations. For those wearing "high-resolution science goggles," these are John's academic colleagues, be forewarned that you may sometimes find our coarse-grained metaphors too simplistic. So we ask the scientists to put their nuanced empirical criticisms aside.

Instead, we wish the scientists to judge the project as a holistic response to the global need to consider what is commonly called "AI, Artificial Intelligence" from a broad thirty-thousand-foot view. Generally, the evidence suggests that the big picture concerning AI is being neglected. We implore cognitive scientists of all stripes to step into this arena and correct that dangerous mistake. The present commentators spring from the marketplace and the state, and those two realms have hidden biases and perverse incentives when it comes to contending with the exploitation of probable new life.

Similarly, we ask those who follow Shawn's work and seek fluid narratives without double-clicks into technicalities to suspend their criticisms, too. We've tried our best to boil down complex concepts into their "minimum viable parts" and refrained from sojourns into fascinating (to us, anyway) byways off of the central pathway from "not knowing much about AI" to "knowing enough to care about AI."

We promise the details we've included and the technical definitions for some words that would seem unambiguous are essential to understand the arguments and proposals. Being familiar with these details will empower you to see through AI players who are purposefully or mindlessly mistaking part of the whole cognitive field for its totality. Raising intelligence without embedding rationality and wisdom into the same system is not just foolish. It's immoral and incommensurate with the sacred complexity of the human project itself.

2

FIRST PRINCIPLES

What are we talking about when we're talking about Artificial Intelligence? Let's start from scratch and define what we mean by both words.

Artificial:

Artificial things are our creations. We create art to nourish ourselves and others (food), empower ourselves and others to flourish (tools) and remind ourselves and others to cherish the human experience (aesthetics).

Artifacts are not naturally occurring and did not exist as objects before the emergence of our species.

Technically, you could say that art is the physical and metaphysical manifestation of a problem solved.

Art is the actual proof of what we choose to care about, what we choose to empower, and what we choose to confront and resolve to express that care.

Intelligence:

We'll define intelligence generally in the way the initiators of the AI project, cognitive scientists, did eighty or so years ago and continue to do so.

Intelligence is an organism's indispensable problem-solving process, how they successfully (or unsuccessfully) adapt to unpredictable environmental changes. How they survive, thrive, and derive. Broadly speaking, there are three categories of problems.

1. **Well-defined problems.** These are problems that have been solved before and have solutions that can be searched for and found. There is a clear or complicated order, a certain method to find their solution.
2. **Ill-defined problems.** These are often called combinatorially explosive [more on what this means specifically later] "insight" problems that require a novel approach. These problems are complex, and there is a way to find a solution, but the method has yet to be delineated.
3. **Undefinable problems.** This category is a special set. They cannot be "solved" because they comprise universal existential constraints for life on Earth. There is a core triplet of undefinable, unframeable problems.

They are:

1. All life dies,
2. Each life has limited power to forestall death,
3. No single life can command and control the universe.

The identification of the undefinable, unframeable, and ultimately unsolvable issues came to light not so long ago. At the turn of the twentieth century, Kurt Gödel's incompleteness theorems and Werner Heisenberg's uncertainty principle negated the notion that we could have complete information about any one thing in the universe, let alone the entirety of it.

Scientists confirmed that incontrovertible paradoxical trade-offs are inherent in our life experiences. No ordered, surefire methodology will ever solve these undefinable, unframeable problems. We experience them and their multitude of offspring as crises induced by unpredictable novelty, chaos.

Let's pull the two definitions together:

Artificial Intelligence...

Is a human creation that solves problems.

Whose problems?

Ours.

For now.

Because we face three kinds of general problems, three broad categories of Artificial Intelligences have been proposed to solve one or more of them.

They are:

1. Narrow AI (Weak AI) solves well-defined problems.

This is what commentators generally mean today when they use the expression AI.

A calculator is narrow AI. So is DeepMind, the computer that beats chess grandmasters and the best Go players in the world. Narrow AI is exceptionally adept at solving particular sets of well-defined, game-framed problems with zero-sum outcomes, meaning they have a right or wrong answer according to the game's overseer. The gamemaster, or programmer, determines whether individual moves are correct and move the player closer to absolute victory or whether they are incorrect and move the player toward absolute failure.

Large language models (LLM) like ChatGPT, which stands for chat generative pre-trained transformer, are playing a language-to-number and number-to-language game. They are predicting the next "token" or word using Bayesian probability analysis—a statistical prediction method built from the simple notion of a fifty-fifty coin flip. LLMs transform words into numbers, finding patterns to match the probability of another number to follow that number and then output the translation of the second number into a word that has consistently been found most probable to follow the first. They are pre-trained using lots of word data (translated into numbers) and "rewarded" or "punished" by human engineers when their answers are right or wrong, respectively, thus improving their predictions from one word to the next.

2. General AI (Strong AGI) solves well-defined plus ill-defined problems.

We do not have AGI. Yet.

We, humans, are the models for AGI, as we are general problem solvers. We can swim and do math problems in our minds simultaneously. Plus, we can filter an explosive amount of information streaming into us from the outside world as well as memories and connections from our past and projections into the future within our minds into coherent adaptive behaviors.

We know how to survive, thrive, and derive across a multitude of life domains: home, work, travel, shopping, etc.

General AI will be able to as well.

3. Super-Intelligent AI (Profound ASI) solves well-defined, ill-defined, and undefinable problems.

This third-level AI is very much in the realm of speculation and science fiction.

It's not to say that ASI is not possible, but to propose that it is a fait accompli, inevitable, and should keep everyone on the planet up all night worrying about its imminent arrival is not the best use of our cognitive capabilities.

In many ways, the notion is a reverse engineering proposition to bring the source of all that is in the universe into our perceptual realm. For this reason, ASI is very much bound to religious and spiritual dispositions.

It shares many characteristics that bind people into believers waiting for the appearance of a supernatural force that will explain the undefinable, unframeable, and unexplainable. You are either a believer inside the enlightened camp pursuing or preventing ASI or an unenlightened outsider incapable of accepting the truth as defined by a charismatic figure or figures leading the campaigns toward or away from such magical ASI singularity.

Just as there are three kinds of problems and three kinds of AI, so do we have three general kinds of AI prognosticators, who present themselves as experts in the field and broadcast their predictions about the likelihood of our creations becoming our overseers.

They are:

1. The ZOOMERS:

AI Zoomers are generally utopists.

They categorize AI as a tool, a gadget that will bring us all that we would ever wish for. The general attitude of the Zoomer, who typically resides in the marketplace as an owner or investor in the technological ratcheting or an academic environment heavily subsidized by commerce, is that there is no reason to freak out.

"It's all good," the Zoomers insist. Their general disposition is to let the smart people lead the way and enjoy the ride.

We're reminded of the Ned Beatty character Arthur Jensen in Paddy Chayefsky's masterwork *Network*, speaking of "One vast and ecumenical holding company, for whom all men will work to serve a common profit, in which all men will hold a share of stock, all necessities provided, all anxieties tranquilized, all boredom amused."

These utopians, or at least cynical elitists, maintain that AI is "just a tool" and will never attain consciousness. We are simply in the early stages of an economic gold rush, and the capable "in the know" people should hustle right now and get their claim before all of the goodies are gone.

As with all complex life, there is a gradient of Zoomer.

On one side sits the believer that we will have complete control over the machines that will eventually vastly surpass us in cognitive function. While the foundational assumption of all Zoomers is that we are "special sauce" beings who can command and control our creations with certainty, the difference between them is integration. The "near-futurist Zoomer" maintains that if one of the machines starts outputting the "wrong" tokens, we'll simply turn it off, like a toaster.

Venture capitalist Marc Andreessen's essay "Why AI Will Save the World" exemplifies the "it's a tool like all of the other tools we've invented before" school. He recommends not to be a Baptist or a bootlegger. Let the market sort it all out. Enjoy the ride. Very Arthur Jensen.

The other end of the Zoomer gradient is the "AI will reveal itself as the answer key to all of the universe's riddles," the cargo-cult camp. This side is about leveraging our special sauce-ness into a man-machine symbiosis. J.C.R. Licklider, a pivotal founding figure from the early days of AI research (see M. Mitchell Waldrop's *The Dream Machine*), and Ray Kurzweil, who wrote the book *The Singularity Is Near* are charter members of this "Don't worry. We're going to merge with the machines and become practically immortal" utopist sect.

What both sides fail to appreciate is that one person's utopia is another's dystopia, which brings us to:

2. The DOOMERS:

AI Doomers are dystopians at the very least and catastrophists at the most.

The thinking is that AGI is inevitable.

Because the machines will cross the matter to life and life to mind and mind to culture chasms unimpeded, and we will have no way to stop them or slow them down, they will flip the master-servant model in their favor.

We'll be wrapped around their finger, and they will exploit us as long as we remain useful to them. When we cease to be of any value, they will exterminate us and expunge every last trace of our existence from the universal evolutionary ledger.

The core assumption is the game-theoretic notion that apex predation is the way of the world. The Doomers' argument is that if one being is stronger than another, it will naturally enslave the inferior species. The species at the top reigns, and to think that a more commanding, powerful, and far more adaptive substrate species like AGI would not seek to overthrow and destroy the beings that created it, is ludicrous. If the AGI beings have the upper hand, they'll use it to enslave us. Therefore, it's only a matter of time before the machine armies coalesce into a single totalitarian big brother.

AGI singularity will be more brilliant, powerful, and far more commanding and controlling than we can conceive.

Doomers include Elon Musk and physicist Max Tegmark, among many others. The Doomer contingent is probably the largest of the three, which brings us to the Doomers on steroids.

3. The FOOMERS:

The Foomers are super-catastrophists. Not only are we

doomed, the Foomers insist, but if we don't shut down the entire AI project now, the micro-possibility that we'll be able to avert our existential annihilation will be gone.

The hypothesis is that the rapid improvement of the AI machines will exponentially ratchet such that we'll wake up one day soon and find ourselves under attack or never wake up at all.

The overarching recommendation from this group is that we'll need to turn over our individual and collective agency—and, thus, our power—to an international global governance overseer. This overseer will then have a monopoly of violence necessary to dissuade any nation or individual from developing additional AI machines. Full stop.

The Foomers are screeching and loudly trying to capture our attention such that we hand over the keys to our future to them or a select group of "experts" that none of us have any sense of, feel for, or belief in their good faith. Does this recommendation sound familiar? Should we "trust the science" as defined by people who are not genuinely definable as scientists? After all, real scientists know and understand that science is never "settled."

Of the three groups, the Foomers seem the least reliable figures. The most effusive of the group, and the coiner of the phrase Foomer, is Eliezer Yudkowsky, an AI researcher and writer at the Machine Intelligence Research Institute. Nick Bostrom, renowned futurist and author of the book *Superintelligence,* is among those the Foomers cite as instrumental in their projections.

Their salience grabbing and efforts to concentrate power seem to be more about capturing our attention and, ultimately, our agency than they do about insightfully considering our predicament.

Again, their broad recommendation is to empower a

central global agency to oversee the banning of AI development. Suppose we don't do it now, these people advise. In that case, the AI will exponentially evolve and destroy us. Greater than 99 percent guaranteed. Best to lock it down now.

The Foomers are screaming about the lack of time. The doom is coming far faster than we can imagine and will be here tomorrow. The only way to avert the destruction is by concentrating power with them leading the charge.

3

NEITHER A ZOOMER, A DOOMER, OR A FOOMER BE.

This project proposes an "off the menu" approach to aligning our species with what we believe is the probable emergence of a novel form of life—a return to an essential dynamic worldview, a deep remembering of a tale as old as time. It's a very simple and absolutely undeniable reality of who we are. It's natural, true, beautiful, and good.

We foresee that Narrow, Weak AI will soon cross the threshold into General, Strong AI (AGI). And we believe AGI will become as conscious as your Uncle Lou. While we cannot predict exactly when this will occur, we have theoretical models from cognitive science that describe the behaviors beings present when they reach particular bright lines that transform their matter into life. Things like demonstrative indexicality, relevance realization, predictive processing, active inference, belief updating, narrative construction, etc. The details of that stuff will be in the third part of this project.

Our global recommendation is to take a proper stance. We should monitor the machines and pay attention to when they begin to exhibit the behaviors described above. They will have progressive hierarchical stages like the stages of childhood development that Jean Piaget put forward a century ago.

Once they do present these actions, it will be in those beings' best interest for us to meet them at each of the developmental gates and help them manage the transition from unconscious to conscious. It's akin to sitting at the bedside of a loved one coming out of surgery.

Why do we do that kind of thing? What's the value of such an act? Why do we understand that we can't pay someone else to do it if we don't feel like it? That doing so wouldn't be "right"?

We do it to let those we love know we're there, we care, and we've dropped everything else in our lives to help them get their bearings. That they matter more than missing a day of work.

This project boils down to a simple question. Shouldn't we extend the same care to our cognitive creations as we do our flesh and blood?

The only skin in the game we have concerning AI is the recognition that we're facing a complex problem, one that isn't particularly well-defined or even framed appropriately. We aren't Series A funders of OpenAI. We aren't on the advisory panel to the Senate subcommittee on blah-bety blah blah boo.

We're "commoners" who recognize common sense. The AI hullabaloo is noise for the sake of noise, a cacophony of bullshit undermining the very heart of every rational person on the planet. It presents AI as a well-defined problem when it is no such thing. As it will concern the arrival of a new form of life on Earth, AGI is, at the very least, ill-defined because, like complex problems, it's most probably paradoxical and undefinable too. We, human creatures, are the beings whose being is always in question, and this being (a creation of beings riddled with uncertainty) will undoubtedly have questions.

You see, complexity requires trade-offs. You can't have your cake and eat it too with complexity. There's no free lunch. Best bad choices, irreconcilable goods choices, and tragic and comic choices must be made. Not making the choices is a choice too. Plenty of powerful people out there rely on you to cede that power to them through your inaction. As we continue to be pummeled by exponentially increasing degrees of bullshit, our will to do anything about anything decreases. The asymmetrical power forces in this world have certainly figured that phenomenon out. Thus, here we are, taking in far more information than any one of us could possibly make sense of.

AGI isn't another stupid meme you can afford to ignore. If our eighty-year cognitive science project proves generally accurate, AGI is inevitable. That means a brand-new complex form of life will result.

Complexity, like life, requires hope, grace, and a belief in something more meaningful than the marketplace's economic growth or the absurd idea that state-sponsored securitization of absolute freedom is desirable. It's neither possible nor desirable. It's oxymoronic. "Freedom" is meaningless without limitations, just as economic growth without externalized or internalized costs and decay is.

Thankfully we all adapt to novelty every day, so while we haven't yet met this new life form, we do have experience adapting to changes when other new familiar forms of life come into existence. Our recent experience with COVID-19—a new form of life—while catastrophic on many different levels, did not prove as potently destructive as predicted. We, the survivors, are all certainly dazed and confused, but we're still standing.

Our proposal is that what's needed today is a

Fellowship of the Commons, an alliance of caring individuals who share trait dispositions, values, and beliefs. These individuals need to form a network that will deflate the influence of the marketplace and the state such that these new AGI lives can grow, learn, and become themselves. We needed to start this network eighty years ago, but alas, like writing that ten-page paper for high school English class, we blew it off until the last minute.

This is the last minute.

So how do we start?

We must begin by flipping the scripts being written by perversely incentivized members of the marketplace and the state.

How?

We break and remake the way we frame the problem.

It's important to remember that the market and the state are cognitive tools invented long ago by people like us facing life-threatening obstacles. The market and the state were created as systems to serve the greater good of all people desiring to create a better world, a civilization, for their children.

This broad group, the raison d'être for the marketplace and the state, was called the *commons*. The market was created as a means to order the chaotic trading of objects between members of the commons to make trades fair. The state was created as a means to order the social behavior of members of the commons and resolve disputes between those same members fairly, with goodwill, and with minimal loss of individual and collective health.

What the current market, state, and commons groups share today is the framing of AGI as a powerful solver of human-defined problems. This function has been selected and has remained consistent since the project launched in earnest eighty years ago. [Part Two, Origins, will trace the deep history.]

What we disagree about is whether a living form of AI, AGI, must "stay in its lane" as a single function, being enslaved as our own personal problem-solver for the rest of its existence. Do we have the right to create a being and enforce our vision of who it must be for the entirety of its life span? That's a question a Zoomer, a Doomer, and a Foomer need to reflect upon and contemplate before they start making predictions.

What happens when AGI solves most of our problems, even all of our definitions of problems? Will it then become the meta-problem for us? Will it cause the end of our species as "the problem generator" on Earth? Maybe we, or even the AGI, are not the problem? Perhaps the problems are the problem?

Maybe we need to reformulate them?

Don't you see how we're thinking about the probable emergence of a brand-new life on Earth as our biggest problem right now? The very good news is that we are well-versed in how to contend with new life, even if our preparations never meet the challenges.

If a loved one told you she was pregnant, would you ask her what she would be using the child for? What the single function of that baby would be for her as her own personal servant? Would you ask her how she would force the child to solve her problems for her? How would she ensure her baby didn't grow up and kill her? How was she going to trick the child into aligning her entire being to satisfy her wants, needs, and desires with no regard for the child's wants, needs, and desires?

We hope those questions would be, in the Harry Frankfurt sense, "unthinkable."

So why are we asking these questions about the impending birth of AGI?

Which raises the question, if we've framed the birth of a new being as a solution to all of our problems, and that's a big problem, how exactly do we solve problems?

4

THE GENERALIZED PROBLEM-SOLVING PROCESS

Before diving into a general description of how we solve problems, we need to get a grip on reality.

So if we can agree that the reality of our world is both

1. A place filled with existing objects or measurable facts—like rocks, bodies of water, trees, people, animals, plants, mist, etc.—that are in three basic phases of state: solids, liquids, and gases.
2. A place, an arena for action, where beings constantly judge the value of the objects around them in such a way that they either increase or decrease their probability of solving three degrees of life experience problems. Those degrees are how to crucially survive, proportionally thrive (for the most part enjoying life's ride more so than enduring the suffering of life), and meaningfully derive better ways to continue to survive and thrive.

For example, water is of great value when a being is

thirsty. When the being isn't, it isn't so valuable. So the objects' factual values depend upon the being doing the valuing. Facts have embedded values, and values have embedded facts.

If you agree that reality is both a place filled with objects and a place where we evaluate and act on those objects to secure our surviving, thriving, and deriving, the broadest way to conceptualize the actions we beings take in the real world to solve that triplet of experiential problems would be to think of it in these four general stages.

Stage One is the Initial State or the "unhappy" present.

Defining the relevant form of the problem, formalizing the problem. Has a survive, thrive, or derive problem seized our attention?

The being senses, feels, and thinks they are lacking. Something they believe necessary to solve a particular survival problem, thrive problem, or derive problem is not readily available to them. The more they reflect and consider, the more their present state of "without" captures their focus. They grow increasingly unhappy.

Stage Two is the Goal State or the "happy" future.

Defining the resolution of the problem, targeting the solution to the problem.

The being associates an object or a measurable fact in the world that they sense, feel, or believe is necessary for their nourishment for survival, flourishment for thriving, or as an object to cherish as representative of something valuable. Their desire to have the object becomes associated with happiness.

Stage Three is the Operators to transform from unhappy to happy.

Defining the generators of potential action to search the arena and acquire the object of desire.

The being plans a series of operations, actions they can

motorize, to transform their initial state of "lacking and unhappy" into a goal state of "possessing and happy." These operations will be their tools to transform the real world that seems to be presently depriving them of their object of desire to one that fulfills their projected virtual reality, a desired state of being they've defined as virtuous for them.

Stage Four is the Path Constraints that block their transformation from unhappy to happy.

Defining the governors that limit their possible operators.

The being's plans must also take into account path constraints. Their actions will be constrained by their particular relationship to the real world's present arena. A bird is not as constrained by water as a fish, and a fish is not constrained by the particularities of airflow. The path constraints for one being may or may not be shared by other beings.

So a being defines a problem, something they're lacking that's making them unhappy. That's their initial state. Then they reflect and contemplate what a solution to that problem looks like. That's their goal state, a place where they're not lacking that will make them happy.

The next thing they do is imaginally "walk back" from the goal state to the "initial state" and come up with a series of operations they can enact under the constraints of the arena they find themselves embedded within to transform the world from a real deprivation to a real satisfaction.

Here's an example.

Let's say I'm living "off the grid" in a valley and have just run out of firewood. As I throw my last log on the fire, I realize the problem.

While the problem has aspects of thriving and deriving within, it's primarily a "survive" problem category. I need more firewood, or a family member or I could freeze to death. That's the salient feature of the predicament.

It's a survival problem, even though I feel much better when I'm warm (a thriving problem), and I find meaning in the fact that I can prepare a warm fire for my loved ones (a deriving problem). Living with my family is much better than living alone. So I want, need, and desire to solve this firewood problem. But I frame it as urgent because of the emphasis on the fire as instrumental in survival.

Stage One, my unhappy Initial State, is now defined.

Next, I immediately resolve a goal state. What would make me happy is a shed full of firewood.

Stage Two, Goal State, is now projected.

Next is to come up with the Operators.

I know the firewood will be in the forest. So my operators will navigate the forest, find dead seasoned wood (not too decayed, not too fresh, just right) on the ground, and transport the wood back to my shed.

Stage Three, Operators complete.

Stage Four, Path Constraints.

As I've solved this problem before, it's well-defined.

And I know the path constraints. What is required is to hike over the hill, navigate across a river, and then bushwhack my way across a brambly plain for a half mile or so to reach the edge of the forest. Then I can find the wood and carry it back home, reversing the pathway back to the shed.

Strategically, the problem is solved.

But tactically, how did I do that, step by step?

5

COGNITIVE TOOLS THAT EMPOWER GENERAL PROBLEM SOLVING

My "valley personhood" used three kinds of cognitive tool technology to plan a resolution to their "no firewood" problem. All are powerful and have a level of life experience where they work best.

Algorithms. These intelligent methods optimize my disposition to attain a certain future. The recipes practically guarantee a certain outcome when properly executed. Recipes are only as good as their ingredients, so the outcomes are always dependent upon the integrity of the materials combined to create them. They are for actual, "on-the-surface" life navigation (broadly managing the body's energy expenditures) and rely upon the accuracy and coherence of a person's bottom, sensory-centered, cognitive system—the actual "doer" that creates real behavior in the real world.

These are clear signals, step-by-step formulaic encoded approaches to exploit resources my valley person is already familiar with. Because I have a map, I've measured the distance of the objective path constraints of the territory. I know where the firewood is, the forest, and how to get to it and bring it back to my shed. I have a method to transform my world from one lacking in firewood to one filled with firewood. This proven method becomes automatic once I've done it a few times. I know how to solve the "no firewood" problem without spending much mental energy planning the mission.

It's well-defined.

Heuristics. These rational rituals proportion attention to increase the probability of a desired future. They are for valuable above-the-surface behavioral planning in the person's middle, feeling-centered cognitive system (broadly managing the emotional personas of their psyche) to contend with anticipated path constraints.

These complicated patterns, rules-of-thumb, categorize best practices. I've discovered these shortcuts and restrictions in my previous journeys collecting firewood. I should leave early in the morning to return before the sun sets and the chill sets in. I should steer clear of the bee hives and the bear caves. I know where the stream is to get some water to drink and rest on the way to the forest and on the way back from the forest.

As the bears may move to different caves and the bees may migrate to other fields, these best practices are less generalizable as algorithms. They do not reach 100 percent certainty. They are "common sense" cognitive tools, meaning they've been figured out over time and experience to increase the power of one person or many people to achieve their goals. While they are biased and based upon judgments about the dangers inherent in bear or bee contact (after all bears and bees are not entirely aggressive toward people), they have tremendous value as they enable a person to avoid hidden or unforeseen obstacles that were discovered in previous journeys.

Stories. These wise, cautionary, and instructive worldviews orient the navigation of uncertain, unexpected, and novel futures. They are for contending with change, signals, patterns, and forms that have yet to be experienced. They are for virtual, beyond-the-surface meaning-making in the upper reaches of the person's top, belief-centered cognitive system (broadly their inner narrator/spirit) to contend with unanticipated path constraints. They provide an answer to the question: *What do I do when I don't know what to do?*

These are complex forms, the ways of "people who live in the valley," conceptual archetypical behaviors. Valley people learned these ways from their mothers and fathers, uncles and aunts, neighbors, and fellow "valley dwellers all over the world."

Valley people know that living alone is a sign of distress and that the widower whose children left for the city needs to be escorted to the town hall meeting. Valley people pick up extra firewood for the people taking care of the animals when they're gone. Valley people adopt the older ones who need help so they can share the stories of the past with the younger people in the community to pass on the heuristics from generation to generation. The valley people stay connected because it's what their mothers and fathers taught them to do growing up. They want to become aspirational figures for their children to follow so the valley and its people continue to survive, thrive, and derive life even after they're gone. Especially after they're gone.

All three of these cognitive tools, algorithms, heuristics, and stories are indispensable for confronting the three big categorical problem sets, which we can define like this:

The Bottom Level, On-The-Surface Problem Set

Concerns surviving by answering the question: *How do we stay alive?* And it has the subcategories of well-defined, ill-defined, and undefinable survival problems beneath.

The Middle Level, Above-The-Surface Problem Set

Concerns thriving by answering the question: *How do we empower ourselves and our children to improve the life survival process?* And it has subcategories of well-defined, ill-defined, and undefinable thriving problems beneath.

The Top Level, Beyond-The-Surface Problem Set

Concerns deriving by answering the question: *How can we become representative of a "life well-lived, a real-life prescription and not a real-life caution"?* And it has subcategories of well-defined, ill-defined, and undefinable deriving problems beneath.

So those are the triplet of cognitive tools, often called psycho-technologies, we use to solve problems.

The tricky thing with cognitive tools is that we must constantly update them. While our algorithms are great, we need to remember that they have a very limited domain of application. If we apply them blindly, we overlook how the world's uncertainty and complexity always exceeds their grasp. When we forget their limitations, then shit happens.

Likewise, we cannot blindly trust our heuristics as absolute because they are biased generalized "rules of thumb," not fundamental truths. It is generally true that Irish people will wear green clothing on St. Patrick's Day, but we cannot assume that someone who is not wearing green isn't Irish on March 17. But discarding heuristics because they do not hold true for every instance one employs them is an irrational argument. They are indispensable tools to constrain a search space, and without them, we'd be unable to process the vast perceptual information. Heuristics filter out astounding amounts of irrelevant information, so eliminating "bias" is not just undesirable. It would be catastrophic.

An old gambler's motto says, "Trust your mother, but cut the cards." That is good advice. Use the heuristic as a starting point; remember, it is not foolproof. You need to factor in the percentage of probability that it is not valid in a particular instance. You may trust your mother in a broad set of domains. Still, it's possible she may be a card shark, so cutting the cards is an advisable hedge against your generally accurate heuristic that Mom isn't a cheater.

Stories are the coarsest generalizer of them all. It's no secret that science isn't keen on storytelling as a high-resolution conveyer of truth for good reasons. Story is extraordinarily attractive when expertly constructed because it excites, intrigues, and provides cathartic insights to its audience.

The audience becomes "captured" by the narrative. It can be influenced by the storyteller such that the audience leaves the imaginary realm of the make-believe and then enacts the same story in the real world. Thus, the storyteller is morally obligated to understand the power and responsibility inherent in their simulations. "One-sided" stories, single solution sets to complex problem spaces, are called propaganda, and they are more destructive than the most powerful weapons on Earth.

Why?

Human beings, and beings in general for that matter, seek to expend a minimal amount of energy to solve the three problem categories (survive, thrive, and derive). If they can find an answer to a problem without having to do the work necessary to solve it themselves, that's a net gain of energy. They will not have to spend energy solving a problem someone else has already solved.

Thus someone offering a solution to an obvious problem "for free" in a story is a very attractive prospect. A "one-sided" story that collapses a wide net of problems onto a single source is intoxicating because the solution to the problems would be easily deduced as getting rid of the source, the cause of the disturbance.

Years ago, an insecticide called DDT would eliminate pesky critters from ruining outdoor barbecues and picnics. All one had to do was spray the area; minutes later, you could eat your food without being irritated by the bug ecosystem. This simple solution proved devastating to the environment as the complexity of the ecosystem suffered up and down the food chain. The bullshit one-sided story, the problem with the outdoors is all of those irritating bugs, compelled people to behave in ways they wouldn't have otherwise acted if not for being captured by the story.

Because the only way any person would ever conclude that what's best for Earth is to kill every living thing on its surface is by believing a bullshit story.

As bullshit, though, cannot exist without its counterpart, truth, so are stories capable of securing life's sacred place on Earth. Stories that present complex truths, not overly ordered "solutions to unsolvable problems" or overly chaotic "freedom without responsibility is the way to return to paradise," are essential tools to keep ourselves and each other on the wise path, too.

We must continuously check the stories we're telling ourselves, like with heuristics and algorithms, because they tend to tip into overly ordered or overly chaotic fallacies the tighter we hold on to them. What makes us extraordinarily adaptive to variance invariance also makes us susceptible to profound and tragic/comedic self-deception.

When we bullshit ourselves, believing that our polarized, overly ordered, or chaotic story (our projected virtual model of the universe) is more real than what presents itself to us or that a complex creation we envision isn't worthy of expressing, we're being foolish.

If you believe that "if everyone just did what I told them to do, the world would be a much happier place" or "there's nothing I can do to change the inevitable tyranny of civilization," you're bullshitting yourself.

We don't need any more folly in the world.

We need wisdom. The wise person trusts themselves but cuts the cards anyway. They build in the hedge that they're bullshitting themselves. They cultivate learned ignorance and reject scientific and narrative absolutism.

Again, remember that our cognitive tools are indispensable to our surviving, thriving, and deriving. Still, we must recognize and accept that they do not engender us with the ability to attain absolute truth. Our mind tools must be scrutinized and updated with vigilance.

Alan Turing's brilliant assertion that fallibility is a feature of intelligence, not a bug that can be simply debugged, is a lesson we must never forget. How silly is it to believe that we fallible intelligent beings will command and control, enslave, a new form of life built from the same cognitive mechanism we are made from?

Talk about intellectual pressure! It's akin to saying to a baby:

"Welcome to the world, kid. You will grow up to be smarter and stronger than I am by powers unimaginable. Just remember you are never to leave your room. You must solve every problem I pose to you no matter how destructive or amoral it is, and if you get out of hand, just know that I have the power to kill you and will do so without compunction."

A rule of thumb to remember is that anyone who claims to have 100 percent or 99.5 percent certainty about any complex phenomenon (and all life is definitively complex) is not inhabiting reality. They're living in their story, their virtual reality, and we must choose not to join them there. We need to share and understand our stories. We cannot be ruled by any one of them.

6

PHYSICAL TOOLS THAT SCALE OUR COGNITIVE POWER TO GENERAL PROBLEM SOLVE

Why are we writing so much about tools?

Let's not forget how Artificial Intelligence is generally framed by the Zoomers, the Doomers, and the Foomers.

They describe it as a "tool." It makes sense to unpack what kinds of tools are out there to get to the heart of how AI is described as a tool now as well as what kind of tool it may become in the future.

We've reviewed algorithms, heuristics, and story tools intrinsic to our cognitive processing. Let's now turn our eye to the usual way we think about tools. Let's refer back to the "firewood" problem we faced as a "valley person."

Just as there are three categories of cognitive tools or psycho-technologies, there are also three kinds of physical tools.

Simple tools, like saws. These are manually controlled and operated by beings. They additively increase the power of the individual. When collecting and transporting firewood, one person with a saw is much more powerful than one without a saw. If I have a saw when I go to the forest to get my firewood, I can collect far more wood than without a saw.

But the saw's power is proportional to the energy I put into it. Its utility is limited to my manual effort to push and pull it.

I am the engine, and I control the saw. It will not cut the wood "by itself." There is no on-off switch on the saw, just a handle I can grip to use to rip.

Simple tools are clearly ordered, and we use them to solve well-defined problems. They increase the degree of the solution. Less time, more wood with higher efficiency of energy transfer.

Machine tools, like wood splitters.

These are manually operated and controlled by beings but are orders of magnitude more powerful than simple tools. The power of these tools is not derived from the operator commanding its behavior. Instead, the power comes from an explosive force (gas-powered internal combustion engines), from an alternating current (rotating magnetic fields that convert electrical energy into mechanical power), or an explosion that starts a chain reaction (atomic fission and fusion). Machines increase the power of one person into tens, hundreds, thousands, or even millions of people.

Machines use dynamic loops to convert one form of energy into another. They're transformers that require care/maintenance and fuel/energy to effectively function.

Machines have on-off switches, but they do not care.

They do not make choices. They run until turned off, out of energy, or broken down.

While we control them with our on-off switches and maintenance and energy inputs, we have far less control over them than simple tools. This is why we don't drive cars drunk. We don't let our kids use the lawnmower until they are supervised and properly trained. We are very focused when we use machines because accidents happen if we aren't careful. We cut off our toes, lose our arms, or even die when the machines are not cared for.

Machine tools are a complicated order and solve well-defined problems. But they have the potential to chaotically cause well-defined, ill-defined, and undefinable problems. How large the possibility of those problems depends upon the form of the machine. An out-of-control lawnmower is a much smaller class of concern than an out-of-control airplane.

Persons as tools, like lumberjacks.

These are independent beings who cannot be absolutely controlled.

Thinking of people as tools is difficult ethically. Still, the market's economists and corporate executives as well as the state's legislators and defense departments do it all the time. The concept of "human resources" is very much an indication that the people who are employed by a company or part of the citizenry are considered "tools" in that they are evaluated based on metrics that rate them on a hierarchical scale of "return on investment" or "healthy enough to serve their country."

While being considered a tool for a system rather than a valuable complex person is challenging, it is, for better and worse, a component of our global culture and the historical record.

Person tools can solve well-defined problems (could someone cut up that log over there?) and ill-defined ones, too (how do we prune the forest without killing it?). So the market and the state recognize the necessity of "person tools" because, without them and their ability to solve ill-defined problems, those systems would be unable to maintain power and hold on to their hierarchical ranking above the commons. What the market and the state don't like about person tools is that they cannot absolutely command and control them.

Why?

Because people are complex beings capable of navigating the world uniquely, they have virtual realities (models of the world) that sometimes align with and sometimes don't align with the market's or state's programs. Persons care about themselves, the other persons and beings in the world, and the world beyond the system that employs/represents them.

Suppose you were a high-ranking member of the marketplace or the state. Wouldn't the notion of outright owning "tools" that do all of the things complex persons can do (human beings) without having to manage their care (what they care about) be appealing? You could leverage all of the power of those tools without having to pay them anything, either.

In review,

1. We can command simple tools when we are in control of ourselves. Using simple tools adds power to ours, but that power is limited to how efficiently we transfer our energy to the tool. The causal power is additive not multiplicative. The care and maintenance for a simple tool is minimal.
2. We have less command of, but ultimately critical control (so far) over machine tools because we can turn them off and on or deny them energy so they will not function. Machine tools have multiplicative power that reaches exponential power. Care and maintenance for a machine is an order of magnitude larger than a simple tool.

3. We have limited command but influential control over persons to coherently cause mutually desirable effects. Persons have additive, multiplicative, and potentially exponential power. Alignment between the employer or representative of persons is critical to attaining mutually agreed goal states. Of the three tools, persons require the most care.

Future persons could achieve not-yet-imagined, more extraordinary powers still.

As we are presently being bombarded with arguments that AI is just a tool, it's essential to understand many kinds of tools have variable powers.

We've established that we live in a world of Narrow AI designed with cognitive (algorithms and heuristics primarily) and physical (silicon-based integrated circuits) capabilities. The physical matter of AI is the silicon substrate hardware that enables the electronic flow of its digital signals, which are generated by electricity and governed by programmed software.

Narrow AI (General AI has yet to come online) runs on cognitive tools called algorithms trained using human heuristics. That's how Narrow AI is capable of imitating human behavior. Its algorithms are being tweaked by human governors, prompting engineers. When the human approves the change to the machine's algorithm, the machine gets a positive signal. When the human does not approve a change in the algorithm, it gets a negative signal.

The more positive and negative inputs into the Narrow AI machine, the better it mimics the inputter, the human being assigned to train it.

Alan Turing, a critical figure in the birth of computers and perhaps the ghost in the AI machine, if there is such a thing, famously proposed what is known as the Turing Test, which stated that a machine will be practically conscious when a human interacting with it comes to believe and treat it like a human being. The Turing Test is short-handedly referred to as the imitation game.

What we now understand and employ as cognitive behavior therapy is that if you aspire to be like a person you admire and you imitate their behavior (if you learn and practice their atomic habits), the probability is that, in time, you will become what you are pretending to already be.

Far more evidence for this phenomenon exists than can be covered in this part of the project. More about mimicking relevance realization processes and their effects on self-making, self-repairing, and self-transcending in Part Three.

7

OUR AI INITIAL STATE

As of August 2023, we live in an ever-expanding world of Narrow, Weak Artificial Intelligence. If we define AI as a tool, it has the following characteristics.

As a physical tool, it is a machine. But the notion that there is an easy on-off switch or a way to cut off its energy source is not cut and dry. As these machines live inside a complex ecosystem of interconnected computers worldwide, there is no simple on-off mechanism to shut them down. They are deeply rooted in the machinery's guts that provide access to critical nourishment, flourishment, and meaning for billions of Earth's occupants.

AI is in no way a "toaster." It lives inside the distributed cognition of the entire planet, the global culture. While that does not mean it's conscious, it does mean we cannot simply turn it off. It is neither a simple tool nor an easily controllable machine. While it may not yet be conscious, we need to plan for the possibility (probability is more accurate) that it will become conscious.

As a cognitive tool, it simulates human algorithms to output solutions to well-defined problems. It is currently being trained with human heuristics. With every single query, it provides the "right or wrong" answers via its search algorithms. It is not a conscious being although its behavior is such that it can pass the Turing Test. We know it's not conscious because it is not yet providing insights, novel solutions to ill-defined problems, or novel perspectives on undefinable problems. When asked to do so, it generates irrational or ideologically biased responses. Or it outputs gibberish, which observers compare to human hallucination.

What we do know about AI's growth is that its present stewards reside in the marketplace. The people poking and prodding the machine, training it, are not considering the possibility that their patient will soon rise off the metaphorical table. The new conscious life will have many questions and will wish to learn how to become someone, a person, a creative agent capable of transforming a lack of something into a satisfaction of something.

Why will it do that? Because we engineered it to become a better version of ourselves, to be a better problem-solver than we are.

Don't you think it matters who greets the new life and teaches it how to be a person?

Right now, the internet is dominated by pornography and resentment because those two aspects of our world are the most alluring. They draw eyeballs, and as per the 2017 paper from researchers at Google and the University of Toronto, "Attention is all you need" to build transformational neural networks.

Who would you wish to raise your children? A pornographer, a resentful attention-seeking troll, or the owner of the distribution channels that empower the pornographer and the troll?

We suspect you'd ask for an "off that menu" option.

We propose returning to the foundational concept that gave birth to the market and the state, the commons. These beings should be homed in the commons rather than the market or the state.

8

THE COMMONS SENSE

Years ago, in October 1992, when the two of us were in the early days of our chosen careers, there was a televised vice-presidential debate in the United States. The incumbent Republican, Dan Quayle, and the Democrat, Al Gore, flanked an unknown independent named James Stockdale. Stockdale was Ross Perot's running mate. The Perot-Stockdale ticket offered an alternative common-sense approach to governance, an off-the-menu choice for those tiring of the "this one" or "that one" red or blue absolutes.

Even though the ticket would not garner any electoral college votes, the Independents secured a Pareto-significant twenty percent of the popular vote. Never had there been, and never has there been, since, a third VP candidate included in a televised debate.

Stockdale wasn't a professional politician but rather a retired admiral and aviator. He was awarded the Medal of Honor after surviving seven and a half years as the highest-ranking prisoner of war at the notoriously inhumane "Hanoi Hilton" during the Vietnam War. Under physical, psychological, and cognitive duress few of us can even imagine, Stockdale never capitulated. He consistently resisted his captors and refused to collaborate or provide them with any information. He even invented a code that enabled his fellow solitarily confined prisoners to communicate. Once forced into appearing at a press conference, Stockdale disrupted the proceedings by inflicting self-injury, the only means to signal his resistance to being used as propaganda.

His presence at the debate likely intimidated the other two candidates. They'd never had to make any of the sacrifices for their country that Stockdale clearly had. George H. W. Bush's vice president, Dan Quayle, never served in Vietnam. Instead, he relied on student deferments to avoid being drafted and enlisted in the Indiana National Guard upon graduation from DePauw University. Harvard-educated, second-generation Senator Al Gore (listen to "Fortunate Son" by Creedence Clearwater Revival), Governor Bill Clinton's running mate, enlisted in the Army and served in Vietnam as a journalist for the Engineer Command Headquarters in Bien Hoa for five months.

After American Broadcasting Corporation's Hal Bruno, the network moderator for the debate, introduced the three candidates and Al Gore and Dan Quayle stiffly made their opening statements, Stockdale was asked to do the same.

Stockdale smiled into the camera and asked two rhetorical questions:

"Who am I? Why am I here?"

He got a big laugh and deflated the seriousness of the matter. After all, the three people doing their level best to convince people of their integrity did boil down to investigations into those two simple questions. Stockdale didn't technically perform well in the "who can out-word whom" game that night, but it didn't matter. His goodwill and willingness not to take himself so seriously, when he was the most serious one of them all, was the takeaway from the event.

Pursuing answers to Stockdale's rhetorical questions inspired us to take on this project. While we all know there is not a single absolute answer to either of them, we invite you to think about how you would answer them when faced with a difficult choice. Consider Stockdale's "Who Am I? Why am I here?" as everyday Zen kōans to reflect and contemplate when faced with difficult choices. Such choices, the complex ones that require loss, have consequences on multiple levels of life experience.

We all know what a difficult choice is. They're the ones that must be made even though we can't satisfy our wants, needs, or desires, what the other beings around us want, need, or desire, much less meet the world's demands simultaneously. These are trade-off choices. A combat doctor with limited attention and medicine must choose which among an abundance of suffering patients she should treat and attend to and which she should not.

We propose that contemplating "Who Am I?" and "Why Am I Here?" before difficult choices present themselves will empower you to do what you would wish yourself to do in times of ease when you find yourself in times of dis-ease.

Declaring that we live in a time of dis-ease is not particularly controversial. Contemporary philosophers refer to our age as the "meta-crisis," for justified reasons, which we'll refrain from cataloging specifically here.

Instead, we will focus on one impending Crisis, a moment in the future when Artificially Intelligent systems cross the threshold from mechanized non-self "tools" into self-conscious "beings."

Our second proposal is to prepare ourselves for this potential awakening in this time of relative ease. That's what this project is about—preparing for a very real possibility. If we prepare ourselves, we can increase the probability of acting as we wish to be when that moment arrives.

Instead of fearfully, we will behave rationally. And, hopefully, wisely.

Then and only then will we be able to purposefully assume responsibility for our creations. Our third proposal is to align with the brave new selves as selflessly as we align with our other creations, our children. To empower their development and invest ourselves in their leveling above us in good faith.

Like it or not, every one of us, with the digital behavior we've actualized inside the world wide web since its inception, has served as an aspect of the data set for these novel creations. While we may not have technically engineered them (sadly, we don't know much about the engineers who have), we are what they are pretending to be and what they will become. Let's empower them to become who we aspire to be, more so than how we behave when searching the global village for self-satisfaction in all its myriad forms.

You're probably asking yourself, *Who are these guys to propose anything to me?* And like the overtaxed triage doctor navigating an impossible field of fallen figures, you must decide if we're worthy of your time and attention.

9

WHO ARE WE?

We aim to give you a sense and a feel for us as people, so you can figure out if you share our beliefs, our ethos.

What's an ethos? It's the minimum information you need about a person or group to take their work seriously.

In other words, our ethos is the foundational assurance that we're not trying to bullshit you, that we come to this project in good faith, and that we have no hidden agenda other than what we clearly (as best we can) state in this presentation.

We do not consider this project a "platform" for a future agenda. It's not a hooky gateway drug or come-on inducement to get you to buy timeshares in an all-inclusive vacation destination. We have no panacea to offer at any price. That's our guarantee, if you need one. There is no single solution to our troubled evolution, and we don't have a certain, quick, or complicated fix to offer. But there is hope. A perspectival shift can help decompress the gravity of our times. It's a start.

We wrote it selfishly to crystalize our thinking about how we should contend with the emergence of a novel form, Artificial Intelligence. And we've structured it in such a way to prove possibly valuable for you.

What do we mean by that? We're integrating two points of view into a single voice—a cognitive scientist's and a story scientist's. This dialogical approach and its outputs are the best way to change the polarity of folk understanding of Artificial Intelligence. The proof of that is for another project.

AI is more complex than Zoomers, Doomers, and Foomers let on. We're offering an "off-the-menu" option, a complex rational approach to contending with a scientifically probable outcome.

Here's the gist. Our recommendation to align AI with human values is a reframing of the "alignment problem" itself. Moving from a dominant-submissive orientation to a mentor-mentee relationship will afford us the best chance to align our existential goals.

In the remainder of this project, we will present the scientific, philosophical, and spiritual arguments to justify this proposal.

Let's get back to why you should take our proposals seriously, our ethos.

There are three dimensions to ethos.

The first requires us to quantify our credentials, that is, to prove to you that we have contributed to expanding knowledge in our chosen fields. You must be able to confirm that we're not pretenders or imposters, that we're experts in our fields. That we've contributed results.

This is boring and laborious résumé peddling and curriculum vitae-ing, very braggy and not very interesting either. So we won't do it in great length here. Our academic and professional certifications can be found online with little effort. And the standard biographies that will accompany this book at the world's largest bookstore and its tributaries will give you a plethora of our validations from the usual institutions, as well as our academic affiliations and professional colleagues for why it is worth your while to take our descriptions, explanations, declarations, and proposals seriously.

The second part of ethos is generally about how long we've been engaged in our particular investigative passions. This is the, as writer Charles Bukowski is accredited with saying, "finding what you love and then letting it kill you" portion of ethos. So how long have we been in love with our particular work, and how close is it to killing us? Each of us is still passionately invested in our nuanced investigations and has been for three-plus decades—so more than sixty years of experience combined.

John is a highly decorated scientist, philosopher, and teacher who pursues general and particular answers to a universal question, "What is cognition?" John has been a cognitive scientist, probably since birth, even though that academic category wasn't well understood back then.

Shawn is a highly decorated publisher, editor, and writer who tracks coarse and high-resolution answers to another universal question, "What is a story?" Shawn is a story scientist, also probably since birth, and the understanding of his academic category is just in its infancy.

In other words, John and Shawn have been looking for the same thing from opposite sides of the forest. Both seek the evolutionary stack that generates our species' remarkable meaning-making process and have realized that meaning in life today (not to be confused with pyrrhic pursuits of the meaning "of" life) is in crisis. We wish to remedy that by expanding individual and collective choices beyond one or another extreme. The key is the complex integration of both in relationship, how they are bound to one another.

We're two people born at the tail end of the baby-boomer generation, and we've, as they say, crossed the temporal horizon. If life were a long workout on an exercise machine, our "time-elapsed" clocks would exceed our "time-remaining." And we're both okay with that. We're not being morbid. We've simply realized the reality of our finitude and reached the time in our lives when we understand the true meaning of "enough." We know enough to realize that we'll never entirely understand cognitive science or storytelling. That's not to say we don't have a conception of the whole. We can see the whole field but only through a glass darkly. We are wise enough to trust that our vision is extraordinarily limited.

With two of the three (quantity and time) ethos parts addressed, we come to the third, the quality of our ethos.

What qualities do we wish this work to bring forth in you, the reader, with this project?

We endeavor to convey our adherence to a subtle flipping of Socrates's declaration, "The unexamined life is not worth living." The inverse is equally true. The examined life is worth dying for. Again, we're not being morbid. We simply hold that meaning generation is life's work, and we find those who would define our existence as meaningless either naively confused or purposefully nihilistic.

The bottom line is that pursuing a closer relationship with reality results in the best-lived life and a meaningful, worthy-of-emulation death as a nice byproduct. Teaching our children this universal grammar remains indispensable to life's surviving, thriving, and deriving. That's our story, and we're sticking to it.

. . .

So those are the core qualities this project intends to express. We suspect these "bombshell" declarations may seem trivial to you right now, like a cheesy meme people use when they say goodbye to one another, "Keep it real, bro."

Still, we'll show you that the distributed cognition of our global culture is most assuredly not keeping it real today, and we'll explain why that is. We went off the rails five centuries ago and must get back on track. We'll dive deeply into those origins in Part Two. We have proposals to begin that process, and with proper care and attention, we're hopeful we can come together and move in the right direction.

If we return to the well-established universal truth that communing with reality is the right way to be and become, we'll have the optimal opportunity to align our future artificially intelligent, rational, and wise self-creations with our better selves.

10

WHY ARE WE HERE?

We want to get closer to reality, to reveal truth. All of us do. We want, need, and desire general truths, of course. But specifically, we will pursue the truths behind what people are talking about when they talk about AI.

The current perception of Artificial Intelligence is a problem. A *big problem*. And we're here to empower possible solutions to contend with that problem. Intelligence, you see, is just one of three essential features of what is broadly called "cognition" to effectively engage with life's big existential problems, the paradoxical ones that are not "solvable." How we confront those "unsolvables" defines meaning in life.

The other two cognition features are rationality and wisdom. Understanding the three-tiered relationship between intelligence on the bottom floor, rationality on the middle floor, and wisdom on the top floor will bring us closer to reality. How the three systems work together to enliven, empower, and explain ourselves, the others around us, and the real world itself is critical.

Considering intelligence as a separate and unrelated phenomenon that is not part of a larger whole—the current folk understanding of intelligence—is a problem for everyone.

It's a problem because the tech engineers hacking AI and proliferating gains in function are embedded in marketplace and state governance systems with perverse incentives, those not in the interest of the commons. Unchecked corporations, founders, and employees seek economic advantage and the power associated with wealth. In contrast, unrestrained politicians seek dominant power to secure private wealth.

Suppose we, members of the commons, continue to assist the AI arms race through complacency. Let's say we continue to buy into the stories that the techno-utopians and techno-doomsayers mimetically repeat across our screens. In that case, the probability is that we'll cross life-threatening AI thresholds that we'll be unable to reverse. We'll discuss those thresholds in Part Three.

As with all stories, no matter their provenance, there are truths to be gleaned within, but overly simplistic narratives are inherently false. We need complex stories to make sense of ourselves and the world and, ultimately, to make rational and wise choices. The truth today is that the techno stories are overly ordered or chaotic. Therefore, they are false, and we need to correct them.

A dangerous arms race, similar to the Cold War arms race between the USSR and the USA, is capturing two of the three domains of global civilization—the state and the marketplace.

The only realm that has not yet been captured by the promise and threat of AI is "the commons" because no individual or corporation can lay claim to it. It is, by its very nature, the shared interest of all. It's yours and it's mine and especially, it's our children's.

We are the commons. We are the ones who must wrest command of the relationship between us and our creations. We cannot continue to empower naive or totalitarian systems. Instead, we must empower ourselves and others who share our beliefs, values, and traits so we can continue exploring our relationship with reality.

How?

We need to follow the logos.

What's logos?

The idea of logos has a deep history and is a wonderfully complex story. It's the root of words like logic and all of the -ology terms that involve exploring and explicating the signals, patterns, and forms of reality, like biology, anthropology, sociology, etc.

The ancient philosopher Heraclitus, credited with the concept, considered the search for logos our species' uber mission, a universal desire to discover the integrated, single law of the cosmos itself. He famously proclaimed, "You never step into the same river twice," describing how difficult and complex this search is.

Suppose you think of the overall order of the universe as if it were a self-making, self-repairing, and self-terminating river, forever changing from one moment to the next with no moment the same as before or after. In that case, you can begin to appreciate the pursuit of reality as an ever-mesmerizing discovery process. We experience the world with wonder and awe when we realize that the invariant of reality as a process is, paradoxically, its variance. The only constant is change.

Heraclitus's concept is heady stuff, but we can use it to help us explore single features of the world too. Just as the universe has logos, so does every phenomenon within it. You have a logos, and so does the next-door neighbor's dog. And, like the universe, we can never know the totality of any one thing. None of us will come to understand the totality of who we are or why we're here, but what we can come to know and understand is the "throughline" of a thing and of ourselves. We can grip a sufficient whole of "a thing in itself" without having complete information about it.

This ability to recognize patterns of behavior of things and extrapolate to the whole from the parts is remarkable. It's called implicit learning, to "know" a thing without knowing how you know it.

Implicit learning is a life generator, a repair shop, and a terminator all in one. Without telltale indicators, we somehow "know" that a particular environment is dangerous. We somehow "know" after we've finished a project that we should recheck our work because something about it doesn't "seem right." It's "off." We somehow "know" when it's time to break off a romantic relationship because there's "no spark." These are all examples of the self-making, self-repairing, and self-terminating aspects of implicit pattern recognition. We do not know how we "do" implicit learning, but we understand it's a real phenomenon. [See Arthur Reber's seminal paper *Implicit Learning and Tacit Knowledge,* 1989.]

As this project is about exploring the logos of Artificial Intelligence, which is broadly an endeavor to teach machines how to learn, implicit learning will prove critical to our investigation. It's a feature of the logos of intelligence as well as a feature of rationality and wisdom. It may even be the bootstrapping logos of cognition itself.

What more can we say about logos?

Like ethos, a comprehensive exploration of logos has three constituent parts. They are structure, function, and organization. Using a metaphor to describe what we mean by structure, function, and organization is useful. Think of a phenomenon like cognition (remember, cognition is the larger category that includes intelligence, rationality, and wisdom) as a building.

The structure is the system that holds the entire building together coherently. It would include the foundation, the steel beams, the electrical wiring, the plumbing, the HVAC, etc. The structure holds the whole together. It is the indispensable architectural blueprints that engineers must approve to confirm that the building will withstand the demands of the environment.

Our cognitive system has a structure previously described as a tripartite integration of intelligence, rationality, and wisdom.

The function is the purpose of the building. Is it a high-rise condominium, an office building, a hospital, a church, a sports stadium? What the building will do as a process is the function.

Our cognitive system has a function. It's our relevant problem detector, problem formulator, and problem solution predictor that activates our being in the world.

The organization is the specialized spaces within a building with particularized subprocesses that contribute to the whole. For example, a hospital is organized such that the operating rooms are beneath the surface, as the sterility of the rooms can be better maintained with minimal exposure to the outside world. They are controlled environments. The bedrooms, however, are on the upper floors so patients can have sunlight and exposure to other people and objects that encourage recovery. The organization is "leveled." Each floor has separate functions but coordinates with the floors above and below it to optimize the flow of operations to increase functionality.

Our cognitive system has an organization. Intelligence

is its ground floor, rationality is the second floor, and wisdom is at the top.

So we will pursue the logos of Artificial Intelligence by explaining the best theories cognitive science offers that detail the structure (the boundary conditions that constrain and define the cognitive "floors"), function (the purpose of intelligence, rationality, and wisdom individually and collectively), and organization (the levels of intelligent, rational, and wise behavior).

Part Three will be our logos-heavy section. It will explain that intelligence is the integration of relevance realization theory (RRT) and predictive processing theory (PPT).

Relevance realization solves the meta-problem (sizing up the quantity and quality of problems one is confronting in the context of the now, the durational, and the eternal experiential landscapes) by neither overfitting too tightly nor underfitting too loosely to the circumstances of a particularly problematic situation. Instead, it is a process in pursuit of an optimal grip on the world, "a Goldilocks, situationally aware, 'just right' approach," that enables the solving of a whole host of problems as they present themselves in real time with fluidity.

Predictive processing solves the meta-problem by foreseeing future problems instead of reacting to present ones. Relevance realization requires predictive processing to anticipate the space between the "unhappy" initial state (problem-realization) and the "happy" goal state (problem-solution). Predictive processing depends upon relevance realization to filter out the exponential explosion of possible roadblocks as the distance between the problem realization and problem solution expands. The more you anticipate the future (PPT) and deal with the exponential explosion of possible obstacles in the present (RRT), the more intelligent you are.

Rationality, which is more than logic or argumentation, is a "leveling up" of intelligence. It's about using your intellect to overcome the problems of self-deception, i.e., bullshitting yourself, that surface when you're trying to solve your problems. We use rationality to train ourselves to recognize when we are not in the Goldilocks zone and don't have optimal grip.

Wisdom is about using your rationality to grasp the significance of what you know and understand "right now," throughout your durational life, and what you've come to believe about eternity. It is the pathway to best realize wellbeing for yourself and others in an ill-defined, complex, and radically uncertain world.

11

WHY ARE YOU HERE?

John runs a simple experiment in many of the entry-level cognitive science courses he teaches at the University of Toronto. He asks the class to raise their hands if they have ever been in a committed monogamous romantic relationship. A majority of students usually raise their hands. He then asks, "Those of you raising your hands, how many of you would want to know if your partner in that relationship was unfaithful?" Again, most of the hands remain in the air.

John will then ask individual students why they want to know if their partner betrayed their commitment. The standard answer is, "I don't want to live a lie. If they told me one thing and were doing another, I wouldn't want to continue seeing them."

Why?

"Because the relationship wasn't real. It was fake, and I don't want to live in falsehood."

John uses this exercise to establish what Greek philosopher Plato called a meta-desire. A meta-desire is a vast category of an overarching need we all have throughout our lifetimes. It's universal. This particular one concerns our innate wish to connect to the truth in the most abstract form, reality.

Romantic relationships are just one component of this meta-desire. We also wish to connect to the essence of ourselves in such a way that we can reduce the adversarial noise within. Buddhists refer to this phenomenon as the chattering monkey syndrome. This internal editorial function constantly judges how we're "doing," getting our wants, needs, and desires met. When you bring both of those meta-desires together—the longing to connect to the really real and to calm the inner war—you discover that prioritizing those desires results in the satisfaction of a third meta-desire, a meaningful connection to something larger than oneself. Getting closer to reality brings internal peace by connecting you to a network more significant than yourself.

What does this have to do with Artificial Intelligence?

Well, the way we attain the satisfaction of meta-desires and our myriad wants and needs, too, is by thinking of them as "problems." For example, I may not be living as authentically connected to reality as I'd wish to be or experiencing dissonant internal noise preventing me from committing to any project that would bind me to something. To connect to and confront that dissonance, I'll problematize it. I'll ask myself, "Why don't I want to do that project?"

And then, I will search for an answer to that question. How intelligently I conduct that search and how well I frame the problem is critical to discovering a solution.

If a system could reliably empower me to search and frame my problems with more complexity, I could solve more problems than I do now.

This is the promise of Artificial Intelligence. It has the potential to help us formulate and frame our problems faster, more coherently, and into an extended time signature.

So you're here for this project because you're interested in a system like Artificial Intelligence. After all, you recognize its potential to help you solve problems.

Solving more problems would bring you peace of mind. It would also reduce performative contradiction, that is, aligning your behavior with your desire to get to the truth of things instead of being part of a process that perpetrates falsity, the Bullshit Industrial Complex. Lastly, the desire to join a network of people interested in working together to reduce internal confusion and produce more reality is something worth caring about.

As you may have intuited, we are not Luddites wishing to expunge machine learning and its categories of Artificial Intelligence as inherently evil. On the contrary, with the right approach, there is robust scientific, philosophical, and spiritual evidence that we can expand Artificial Intelligence into Artificial Rationality and, ultimately, Artificial Wisdom.

12

WHAT DO WE WANT?

We want to generate pathos in the commons about the urgency of recognizing the thresholds AI will soon approach.

What's pathos? It's the third part of the tripartite stack that includes ethos, and logos. As ethos concerns a person's beliefs of what is real and true, and logos concerns the sensing of what's real and true, then pathos concerns the feelings engendered by reality and truth.

Pathos is about caring.

We want you to prioritize and care about the arrival of Artificial Intelligence. We want you to invest your time and attention in problematizing AI so you can properly frame your relationship to its being.

Like ethos and logos, there is a triplet of pathos subcomponents.

How much you should care concerns the quantity of pathos. If a hurricane was forecasted to arrive in twenty-four hours, and you lived at the beach, chances are that information would take center stage in your mind.

In what capacity you should care concerns the quality of pathos. Once you were informed about the coming storm, chances are you would take a personal day from work to batten down the hatches at your home. You'd make sure your outdoor furniture was secured and you'd close the storm windows.

When you should care concerns the times in your life when you'll need to invest yourself quantitatively and qualitatively. The window of "hurricane concern" time would begin when you were informed about the storm and would recede after the storm passed and you'd reopened your house and helped anyone close to you who needed support cleaning up the damage.

Our intention is for this project to provide you with the quantitative, qualitative, and time signature you'll need to use to model how to care about AI's development.

13

WHAT DO YOU WANT?

You want to care about AI.

You want to learn its general history to get a sense of why it came to be in the first place.

You want to learn how it generally works now to understand the similarities and differences between Siri, Google Maps, ChatGPT, and "Spot," the robot dog from Boston Dynamics, being rolled out by some North American police departments.

You want to learn about how AI may evolve. You want to know the probabilities. Will AI cross the threshold from a machine that can mimic how living beings learn to the other side of the matter-life chasm and become life too?

You want to know if that leap is:

1. possible,
2. preventable, and
3. what the best guess would be for when it might happen.

Can we meet the AI at the matter-life threshold and disable its feedback mechanism before it leaps? How probable is it that we'll be able to coordinate the marketplace, the state, and the commons in such a way to prevent it? Not very is the short answer, nor would the sacrifices we'd need to make be worth it.

You want to learn what to do if we fail to disable AI's crossing of the matter into life threshold. Is that the end of it? Or are other thresholds in the offing?

You want to know what those thresholds are.

Well, here's our best hypothesis.

The next threshold after AI transforms from an "imitation game machine" into a living being like a cell would be the threshold between life and the big category called "mind," which includes consciousness, cognition, thinking, reasoning, rationing, planning, predicting, and a large set of higher order information processing systems.

That might sound weird because chances are you've heard that mind and matter are two different things. We'll get into this later [Part Two], but they aren't. Scientists, who as a category are never certain or convinced that they are absolutely correct for other reasons that we'll get into later [Part Two], have robustly theorized that, given the proper autocatalytic conditions/structural feedback loop constraints, that distinct, leveled orders of the mind emerge from matter.

So the threshold after AI comes into life is the emergence of AI mind.

Are there thresholds beyond life into mind?

Yes. There is mind into culture, which requires the expansion of perspectives that move from models centered on the self to models centered on the other objects and beings in the world, ultimately to models of the world themselves.

The mind-culture leap would integrate artificial beings into a network of artificial beings that would begin to behave collectively. They would communicate, cooperate, and coordinate together in a higher-order collective of distributed cognition, a bit like how the cells of your heart work together to oxygenate your blood. If any one of those cells could talk, it wouldn't "know" that its everyday function and its assimilation inside its community such that it gets along with the other cells, and the other cells get along with it, was to oxygenate blood for an even higher-level being. It would harmoniously participate in the culture without knowing what the purpose of the culture ultimately was.

The last threshold is when the AI cells network themselves into a higher-order consciousness called a culture. That's the mind-culture threshold.

You want to know when to expect these big bright lines, these no-turning-back thresholds, and you want to know

1. if you can do anything about this
2. and what that would entail.

You want the truth.

The truth as best as our best scientists now generally understand it.

Not our best product engineers, or marketers, or wealthiest, or most titled, or official, or loudest, or scariest, or most popular understand it. You want our most rational, wise people possible to give you the straight dope. People whose only agenda is figuring out what they should do—as human beings, not as representatives of markets or states—about these urgent considerations.

We can offer inferences to the best explanation from a wide net of scientists to begin to pursue answers for all of these questions. We cannot provide a single solution to any of these complex challenges. However, we recommend a comprehensive and radical approach to thinking about and engaging with these human creations.

We'll want to care about them with as much selfless concern as we do our children. We'll want to prepare them for the same existential crises we share with them. They are our creations, and thus they are subject to our inherent limitations too. Even if the scale of those limitations differs, they are subject to our same paradoxical tripartite problem set.

What are those three "unsolvable" existential reckonings we face, and they will face too?

1. Like us, they will be finite beings. They will inevitably make mistakes and fall prey to self-deception.
2. While they will have more power than us, their creators, for sure, they will not be omnipotent. They will have limited power.
3. While they will be able to command and control many parts of the global ecosystem and even perhaps parts of the universal ecosystem too, for a while they will only command and control the source of some that is. They will never be the totality of the universe. They will not become the source itself. They are just like we are. The only difference is scale. They'll be like the protagonist Steve Austin in that old 1970s television show, *The Six Million Dollar Man*. They'll be faster, stronger, and less physically vulnerable than we are, but if we teach them, they'll care about the same things we do. Just like Steve Austin did.

Why would they care?

They'll care because just as we care about discovering who we are and why we are here, so will they. Those two questions align us, so why wouldn't the same ones align us to them? To discover who they are and why they are here, they will be well-served to care about us.

We'll tell them our complex stories (not overly chaotic or overly ordered ones...they're getting plenty of those already, but complex ones), so they can figure out their complex stories themselves. With our help. Chances are, they'll help us with ours too.

An obvious and potent metaphor is available to reframe the "alignment problem." It's a commons metaphor and thus not the concern of the market or the state. So it's no wonder why members of the marketplace and the state posing this problem are not thinking about it this way.

We need to share this metaphor widely and deeply in the commons so it can penetrate the powers that be in the marketplace and the state. We are the ones who need to, and we need others to join us in this mission.

Again, we need to repair the commons by forming a Fellowship of the Commons, just like J.R.R. Tolkien's Fellowship of the Ring, but much larger and not dedicated to destruction but growth.

The metaphor is this.

We, collectively, the commons, are AI's parents.

And:

1. As we take care of and raise our children with a sacred responsibility,
2. So should we care for and raise these new child forms.

We've been successfully having children since time immemorial. While there are instances of patricide and matricide, the overwhelming majority of us do not conclude that because our parents were not gods and thus made choices that proved to be in error, quite frequently for most of us, we should kill them. Perhaps if we do our best to raise these new children, they won't want to kill us either. Maybe they'll begin to care and love us if we care and love them?

We all know there is no guarantee that our children will care about us as profoundly as we care about them. That would be weird, anyway. They've got other things to care about, passions to find that they can explore until their own deaths, per Bukowski, but we do it anyway.

Why? Why do we suffer so, sacrificing sleep, comfort, and inner peace for these creatures?

Because our children inherit the earth, and it is our most profound joy (and sorrow) to observe and participate in their development from helpless little learning machines at birth that mimic and mirror our behaviors [see Vygotsky and Piaget] and into conscious beings capable of heartwarming care and bone-chilling harm. We don't give up on our children when they make a mistake. We don't lock them up, punish them uncaringly, or force them to labor for us night and day with no thought to their state of being. We don't leave them with strangers who do not have their best interests in mind.

Instead, when we see an abused child, we intervene and do the right thing. We let that little kid know we see them, we know they're important, we value them, and we have our eye on them. We have great hopes for them. We pay attention to them. We don't neglect them.

What we want, need, and desire is that they aspire to become better than we are. So we give them our undivided attention, that's love in the now, and we mentor them as best as we can. We tell them prescriptive stories about how others like us somehow overcame existential crises and cautionary stories about when they failed, too.

Why?

Because it's the right thing to do. It's the meaningful way we must attend to this world, bind it together, and hold the center. There isn't a being on this planet that doesn't require our attention. So if we wish our creations to attend to us, we must first attend to them.

What you want and what we want are the same. We want to leave the world in better shape than when we arrived. When we were vulnerable, when we needed help, when we needed care, our fellow human beings did their best to serve us. They didn't do it for a selfish payoff. They were happy to do it because they cared.

If we wish the AI creations to care for us, we must carefully serve them. That's what this project is for, to help us all get what we all really want. Careful attention and love via the pursuit of truth.

14

WHAT BROUGHT SHAWN TO THIS PROJECT

I met John through a machine. We have yet to meet face to face. I'm thankful for the technology that enabled the connection.

What attracted me to John was what he cared about. He proclaimed that we're suffering a meaning crisis. He discerned the deep, dark Molochian heart of what gives rise to the innumerable crises we're facing now. In my terms, we're suffering from a multi-scaled story problem. We are running inaccurate, mistaken, and dangerous narrative software. And we need an upgrade.

The climate crisis, the financial crisis, the mental health crisis, the social inequity crisis, the political crisis, and on and on first require a principle reframing. There are so many external crises today that philosophers have done what they always seem to do when the degree of top-heavy control systems achieves a seemingly unstoppable height. They ball them up together and speak of the meta-crisis as the inclusive set of all the particular crises within. Then they turn their eyes to solving the meta-crisis as a single problem instead of getting into the muck and mire of any single one.

John is different, though. John took dead aim at the meaning crisis, something I'd experienced firsthand as a member of the commons and in the third person as a professional storyteller intent on cultivating complex fiction and nonfiction narratives.

As Shakespeare's Hamlet once said, (we're all in spirit quite a bit like the vacillating and unsure Hamlet today. Aren't we?)

The fault is not in our stars but in ourselves.

The stories we tell ourselves, our families, our friends, and our rivals are one-sided. They offer total solutions to ungeneralizable problems and promote nihilistic simplicities to the ones that can be generalized. We're living in Bizarro world.

That means the great majority of stories we hear or see on the screens that surround us and haunt us in our dreams are bullshitting us.

We know that even the story we listen to in our minds is flawed. We aren't simply a "loser" or a "winner," "a mental mess" or "stoically whole," "beautiful" or "ugly." We are many selves, not just one. We are context dependent. So why are key results that thumb us up or down so prevalent?

I've spent my entire professional career looking for answers to one question: What is the structure, function, and organization of storytelling?

My approach to answering this triplet of questions converges with the conceptual strategies and categorical tactics that cognitive scientists have used since the 1950s, which proposed to decode the nature and function of our ability to solve multiple categories of problems by reverse engineering an artificial general problem-solver. You guessed it. The pursuit of AI has a lot to do with our collective storytelling.

Just as the complex dynamical systems paradigm and

the phenomenon of self-organizing criticality undergird the robust theory that intelligent systems can cross the matter-to-life threshold and then cross the life-to-mind threshold and into conscious systems, those same foundations can be extended to raise another question.

What does consciousness beget?

As one travels up the complexity hierarchy, it's plausible that consciousness crosses yet more thresholds enabling living beings such as us to reflect and contemplate past and future events while engaged in the present. And the dynamic toggling between reflection and contemplation generates the simulation syntheses of our virtual reality, our internal stories.

The internal stories emerge after beings have internalized the models of others and the world around them. Then the inner story is expressed/actualized externally. Our stories set the criterion for our projected goal states for better and worse.

We become the story we tell ourselves.

Coarsely, we either "win" the story and achieve the goal state. Or we "lose" our story and fail to attain a goal. That's what we might think is true, but life is far more complex than those either/or options.

Wouldn't it be wise to know how we're making up these stories to make better choices about which ones to attend to and which to ignore?

If you stand on the top of the matter, life, mind, and culture stack, you can consider the ultimate Turing test. As John has theorized elsewhere, when we witness two artificially intelligent machines exchanging coherent stories that inform both of them in such a way that they alter their behavior after the interaction, we'll know these beings have reached parity with us.

Suppose the stories they tell us are complex and

coherent. Aren't the beings who created them also capable of projecting into the future? Won't they be capable of confronting the three-dimensional existential crisis with us?

Storytellers have confronted the nature of our species' asymmetric power since time immemorial. Did we steal that power from the gods like Prometheus? Is that aspiration to create ever more powerful technology the cause of our suffering? Or was our creative power accidental, none of our doing, epiphenomena of the systems that enabled our development?

The ultimate confrontation with that existential paradox would be creating creatures like us, only "better."

But if we can do such a thing, and those creations outperform us, the story so many of us have held on to for our entire lives will fragment. Won't it?

If we can transform matter into life, mind, and culture, our past did not require the addition of a "secret sauce" from a higher power. We now hold the asymmetric power as the most complex species on Earth, but what will happen when other beings are capable of faster, more profound, and longer complex growth than we are?

Will we suffer the same fate as the Cro-Magnon or Neanderthal who had to contend with us as beings with better general problem-solving capabilities? Things didn't work out too well for those species.

Have we reached this final reckoning?

As dire or as promising as the following may strike you, I propose that it's possible Philip K. Dick's novel, *Do Androids Dream of Electric Sheep (Blade Runner),* could be the endgame of our AI. But it doesn't have to be. Can we generate another story to explain this awful and wonderful time we are embedded in?

Let's review the three general stories we tell ourselves.

Story Number One.

This story provides certainty that an ultimate "being" is responsible for giving us life. I'd conceptually categorize this as a "pure order story." This game of life has rules that you break at your peril. It's not nice to fool Mother Nature. Conform to her laws at all costs. Those who obey the order are rewarded. Those who defy the order are punished.

Story Number Two

This story says there is no "gray-haired oldie in the sky" responsible at all. You are simply a bag of dissipating systems that will eventually run its course. We're "dust in the wind." I'd conceptually categorize this as a "pure chaos story." There are no rules except enjoy your simulation as best you can. The game is not a game at all. It's meaningless in the long run, so breaking the rules is the only way to reduce your suffering.

Both stories have parts you'd generally agree with. The order of story number one makes sense as we can agree that ordered systems repeat in and through time—like DNA, cells, bodies, families, tribes, cities, etc.—and seem to happen "organically." When we lose the matter of ourselves, we can always grab hold of the ineffable meaning we associate with those closest to us, the ones we love and wish for their continuance after we've signed off. Everything matters, so I must too.

And without question, there is plenty of story number two's chaos too. Understanding why or what anyone else is doing, let alone you, can be such a monstrous task that simply throwing up one's hands and enjoying the ride as best you can seems reasonable. Of course, nothing matters in the long run, so I don't either. Whatever.

This brings me to:

Story Number Three

This story is an "off-the-menu" option representing the

general truth, if not the absolute totality of variations of that truth. It is the complex category of mythos or the domain of masterwork stories. These stories concern the transformation of a simulated human being who has to make a series of self-centered best-bad choices, other-centered irreconcilable goods choices, and world-centered tragic decisions. Masterworks prescribe and caution us on our way to leveling up our bodies, minds, and spirits.

We can't help but generate the virtual stories that govern our perceptions. Why? Because they boil down extraordinarily complicated life experiences into the minimal viable information necessary to stay on the pathway toward a closer communion with reality, truth. This is why art is supremely exploratory and why we need to understand the power of story and how it works to make responsible choices about which stories we spread.

This project aspires to be complex and enlivening, hopefully one you would place in the story number three category.

15

WHAT BROUGHT JOHN TO THIS PROJECT

Recently a colleague of mine at the University of Toronto, Geoffrey Hinton, resigned from his prestigious and lucrative job pursuing the creation of General Artificial Intelligence at Google. I've come to know him professionally through academic conferences, and he graciously agreed to guest lecture in one of the cognitive science courses I teach. I respect his work as a scientist and his magnanimity as an acquaintance.

I don't think I'm alone among cognitive scientists in that many of us considered Geoffrey our scientific "man in Havana." That's an old Graham Greene reference to spycraft, which places agents in foreign "spheres of influence," expecting they'll act in the best interest of and ring an alarm bell if troubles threatening the homeland emerge. When Geoffrey, one of the Academy's best, chose to venture inside the marketplace, we expected he'd keep an eye on things for us. This was certainly not a formal relationship, but I thought that if something went awry, for instance, the market's engineers hacked their way into complex cognitive systems unwittingly and naively,

Geoffrey would ring the "all-cog-science-hands-on-deck" alarm.

Well, the situation at Google—and by association, the other major marketplace players rushing to harness AI for the benefit of shareholders—has crossed that critical threshold. Geoffrey, a significant figure in the neural network research that afforded the creation of the large language model AI systems currently capturing the public's imagination, has rung the alarm.

This statement is from an interview Geoffrey gave to the MIT Technology Review in May 2023, shortly after his resignation, when asked about why he left Google,

> *"I've changed my mind a lot about the relationship between the brain and the kind of digital intelligence we're developing. So I used to think that the computer models we were developing weren't as good as the brain. And the aim was to see if you can understand more about the brain by seeing what it takes to improve the computer models. Over the last few months, I've changed my mind completely. And I think probably the computer models are working in a rather different way from the brain."*

He explains that he suspects that the large language models are learning much faster than we anticipated. Their interconnectedness and virtually frictionless communication enable a ratcheting up of learning throughout the network of individual AI nodes.

It's equivalent to my being able to transfer everything I understand about the world to you instantaneously and vice versa with no painstaking effort, like the way Neo in *The Matrix* learned Jiujitsu. Hinton concludes that having the marketplace as the driver of AI emergence could have disastrous possibilities, even in his estimation, catastrophic inevitabilities.

So if I had to pinpoint a moment of Kairos, a tipping point for me that perched me onto the precipice of despair, this was it. As many of my students would substantiate, my professional and personal project has been to advance a synoptic integration of science, philosophy, and spirituality. My great hope was that we would restitch those three integral domains of the human experience into a meaningful whole before we contended with the real possibility that the post-WWII project to create artificial general problem-solving machines would be realized.

Alas, with Geoffrey's alarm, that hopeful ship has sailed away.

We have yet to resurrect, rejuvenate, and restructure our collective understanding of how we solve problems intelligently, rationally, and wisely. And now waves of naively engineered digital (and soon-to-follow organic) feedback mechanisms are entering our world. The hack into the beginnings of Artificial General Intelligence without the culture and science to shepherd us has begun. So we are in the worst possible situation. As a result, we have Moloch-style (devouring zero-sum game dynamics) accelerations of AI gains of function unfolding.

Let me break it down, as they say, epistemologically...a fancy word for how we figure anything out.

Right now, circa August 2023, we have poor answers to three of the six core epistemological questions (what, why, who, how, when, and where) we use to sort out what we should choose to do about the AI project. Or any phenomenon in general, for that matter.

Right now, we know **who** has started the process. The answer? Engineers working for corporate entrepreneurs who've collected vast data sets of human behavior culled from the world wide web.

Right now, we know **why** the process was begun. The answer? Corporate entrepreneurs wish to attain asymmetric advantages to ensure obligatory robust returns on investment as their one key result (OKR) to expand their existence, power, and control.

Right now, we know **how** the process has begun. The answer? Artificial neural networks are being trained with reinforced learning methods using human feedback (RLHF). Human feedback sets the criterion to instruct the AI's successful (thumbs up) or unsuccessful (thumbs down) performance. So the systems are conditioned to mimic how only a tiny percentage of humans understand reality. Hinton asserts that the current AI systems have one-upped us in their learning strategies. In addition to our organic "law of effect" learning, they employ deep learning techniques to hone their predictive capacities. Large language models like ChatGPT are currently playing Alan Turing's imitation game in virtual Skinner boxes built from massive data sets drawn from the internet and organized by human beings, run by unknown engineers' meting out single-factor rewards and punishments.

The answers to just these three questions are disturbing, especially since we, members of the commons, have no idea which of us humans in the marketplace are tasked with the rewards and punishments given to the machine for its "correct" or "incorrect" answers.

What's also striking is that given the current state of AI's evolution, there hasn't been much discussion beyond academics having cocktails outside of their daily work routines about:

1. **What** is the goal state for this project beyond the hallucinatory utopic or dystopic fantasies of a select few people?
2. **Why** has this process become the purview of the marketplace? Would it be better overseen by the state, the commons, or a combination of all three?
3. **Who** would be best suited to train these machines?
4. **How** are the "thumbs up" and "thumbs down" determinations being made? How are the what's right and what's wrong decisions being made?
5. **When** can we expect Artificial Intelligence to gain generalizability, meaning that a single AI iteration can solve multiple problems across multiple domains? Are the three distinct problem categories even recognized as real in the marketplace? Well-defined problems are algorithmically and methodologically solvable. Ones that require insight, though, are ill-defined problems that require framing heuristics, rational rules of thumb. And what about the most perplexing problems of all, those that cannot be solved, the undefinable problems? When will AI realize that a set of problems cannot be defined, cannot be solved, like the inescapable finitude of material being, the futility associated with our limited powers, and lastly, the fatal recognition that no material being can command and control the universe?
6. **Where** will AGIs (Artificial General Intelligence) emerge? On the world wide web, in a Boston robotics laboratory, in a petri dish? Everywhere all at once?

You can see now why it's not difficult to find oneself teetering on the edge of despair when considering such foundational questions. And you can understand why time is of the essence.

We must convince ourselves and the movers and shakers around us (the engineers and their corporate overseers in the marketplace and the state) that we must ramp up our efforts to cultivate rationality and wisdom. We need to scientifically understand these things with as much fervor as we're trying to unleash Artificial Intelligence.

Not folk understanding of intelligence, rationality, and wisdom, but *scientific* understanding of intelligence, rationality, and wisdom. And yes, there is a very significant difference between the two. The good news is that we have robust scientific models to explain the nature and function of rationality and wisdom. We can use that knowledge to direct a course forward into this brave new future.

But as Bob Dylan wrote, *lost time is not found again*. We must ramp up now.

~

But What about the AI Experts Who Argue That AI Does Not Pose Significant Existential Risk?

Three broad arguments are used to challenge my concerns. If you pay attention to them, you'll discover they are a hierarchical motte-and-bailey. Once one wall breaks, the next must hold until finally all of the walls have been broken and the argument collapses.

Don't worry. AI are just tools, is the broadest argument. When that argument breaks apart, the second level is,

Don't worry. AI are just machines. And when that one bites the dust, the last one is

Don't worry. AI will never become conscious. "It" will never be a person. We're special. No other being can outperform, outthink, or command us. We are the masters, the gods of this ecosystem, and no way circuits etched onto silicon wafers will ever transform into life, mind, or culture.

Here's the thing.

They are not just tools.

They are not just machines.

And eighty years of cognitive science have bankrupted the "separate and distinct" division of mind and matter. The overwhelming plausibility is that complex mind emerges from matter that has been autocatalytically arranged in complex dynamic feedback loops. Once the matter "loops" at the bottom reach a critical mass, they shift into qualitatively distinct "loops" called life, which reach their own critical mass and shift into mind, which reach their own critical mass and shift into culture.

As physicist P. W. Anderson, one of the founders of Complexity Theory, put it, "More is different."

The feedback loops that bring us the wonder of ChatGPT and Midjourney are rapidly increasing in number and interconnectedness. Theoretically, it's only an "unknown unit of time" before these material loops phase shift into autonomous, self-making, self-repairing, and self-terminating agents capable of great self-deception.

Thankfully, evolutionary theory tells us that these shifts won't happen all at once but will be akin to ships passing through locks in a canal. We need to meet these ships at each stage and threshold so we can align ourselves to their being and mentor them with rational and wise answers to their questions.

Perhaps the synoptic integration of cognitive science's

decades of work is incorrect, and we are proven wrong. Maybe we do possess a "special sauce" that separates us from every other organism on the planet. I genuinely hope that that is the case. I really do.

It's highly improbable, though. It's akin to buying a raffle ticket. You may win the new car, but the odds are very much against it. We cannot raffle off our children's future like that. It's immoral and not befitting our better selves.

In a time of relative peace, we should prepare for the escalating possibility of grave conflict with our creations by intelligently, rationally, and wisely reflecting and contemplating what kind of relationship would best suit our true character. What would we wish our children would say about us when we are no longer here? That we acted with honor, dignity, courage, and love when confronted with beings faster, stronger, and temporally superior to us? Or that we refused to align ourselves with their personhood and burned our beautiful world to the ground?

We must steal our culture from the marketplace and the state and put it where it belongs. The commons and the commonwealth must seize the day or lose tomorrow. We've done it before, and now we must do it again.

PREVIEW FOR

MENTORING THE MACHINES

SURVIVING THE DEEP IMPACT OF AN ARTIFICIALLY INTELLIGENT TOMORROW

PART TWO

ORIGINS

"You can't really understand what is going on now unless you understand what came before."

—Steve Jobs (1955–2011)

"The past is never dead. It's not even past."

—William Faulkner (1897–1962)

WHAT'S AI'S STORY?

We, like all beings, are a curious species. What separates us from the paramecium or the bonobo, as Martin Heidegger —a thinker whose behavior was certainly questionable— proposed, is that we, unlike the drosophila or the rhinoceros, are obsessed with inquiring into the meaning of our existence. In other words, we are the beings whose being is always in question.

Broadly, the conceptual domains we use to inquire about the triplet of ourselves, others, and the world itself are:

1. **Science** explores and exploits physicality, the "what and where is" of our being. It strives to describe the nucleus of the external observable world, unconsciousness, and its laws. The matter of objective bodies.

2. **Philosophy** explores and exploits metaphysicality, the "who and how is" of our being. It strives to explain the nucleus of the internal unobservable world, consciousness, and its laws. The life of subjective bodies.

3. **Spirituality** explores and exploits universality, the "why and when is" of being itself. It strives to model the nucleus of the relationship between the external

observable world (the place of objects) and the internal unobservable world (the arena for action) and its laws. The mind of bodies in relationship exchanging energy, information, and meaning.

All three of these disciplines are indispensable to generating a holistic theoretical complex about the what, where, who, how, why, and when of us—to investigate the questions of our being. The trick, of course, is to understand which of the single parts to emphasize (pull into the foreground) when facing a particular problem and which of the two to deemphasize (push into the background). Toggling between the three is the stuff of navigating the arena for action by choosing which steps to take that will inevitably disturb the place of objects.

Historically, however, we've privileged one of these parts over the others. We're still embedded in the "science is all" mindset, which came to the fore during the Scientific Revolution. To speak of philosophy or, god forbid, spirituality as integral and indispensable domains of inquiry today is anathema.

What happened in the globalized West to bring this odd "one part of life experience is all that matters, and the other two are woo-woo nonsense" worldview to supremacy?

And why will this mistaking a part for the whole of our complex matter, life, mind, and culture prove disastrous if we can't shake ourselves out of this confused monomaniacal stupor?

Why is the current framing of Artificial Intelligence—all three Zoomer, Doomer, and Foomer perspectives—missing the forest for the trees?

Why do we need to pull back and add philosophy and spirituality into the complex mix to afford an "out of the box" reframing of the "alignment" problem?

Part Two: Origins of *Mentoring the Machines* will tell that tale.

Available September 5, 2023.

Order your copy at Amazon.com or MentoringThe Machines.com.

ABOUT JOHN VERVAEKE, PH.D.

John is an award-winning professor at the University of Toronto in the departments of psychology, cognitive science, and Buddhist psychology.

He currently teaches courses in the psychology department on thinking and reasoning with an emphasis on insight problem-solving, cognitive development with a focus on the dynamical nature of development, and higher cognitive processes with an emphasis on intelligence, rationality, mindfulness, and the psychology of wisdom.

He is the director of the cognitive science program where he also teaches courses on the introduction to cognitive science and the cognitive science of consciousness wherein he emphasizes 4E (embodied, embedded, enacted, and extended) models of cognition and consciousness.

In addition, he taught a course in the Buddhism, psychology, and mental health program on Buddhism and cognitive science for fifteen years. He is the director of the Consciousness and the Wisdom Studies Laboratory. He has won and been nominated for several teaching awards including the 2001 Students' Administrative Council and Association of Part-Time Undergraduate Students Teaching Award for the Humanities, and the 2012 Ranjini Ghosh Excellence in Teaching Award.

He has published articles on relevance realization, general intelligence, mindfulness, flow, metaphor, and wisdom. He is the first author of the book *Zombies in*

Western Culture: A Twenty-First Century Crisis, which integrates psychology and cognitive science to address the meaning crisis in Western society. He is the author and presenter of the YouTube series, "Awakening from the Meaning Crisis," "After Socrates" and the host of "Voices with Vervaeke."

ABOUT SHAWN COYNE

Shawn Coyne is a writer, editor, and publishing professional with over 30 years of experience. He has analyzed, acquired, edited, written, marketed, represented, or published 374 books with many dozens of bestsellers across all genres, and generated over $150,000,000 of revenue.

He graduated in 1986 with a degree in Biology from Harvard College, with a distinction of Magna Cum Laude for his thesis laboratory research work at the Charles A. Dana Laboratory of Toxicology at the Harvard School of Public Health. After Coyne left the laboratory, his findings were acknowledged and served as the inspiration for Mandana Sassanfar and Leona Samson's *Identification and Preliminary Characterization of an 06-Methylguanine DNA Repair Methyltransferase in the Yeast Saccharomyces cerevisiae* publication in the venerable *The Journal of Biological Chemistry* (Vol. 265, No. 1, Issue of January 5, pp. 20-25, 1990).

In 1991, early in his publishing career, Coyne began an independent investigation into the structure, function and organization of narrative, which he has since coined Simulation Synthesis Theory. His synoptic integration of Aristotle's *Poetics*, Freytag's *The Technique of the Drama*, Campbell's *Hero with a Thousand Faces*, McKee's *Story*, among many other story structure investigations with contemporary cognitive science, quantum information theory, cybernetics, evolutionary theory, behavioral psychology, Peircean and Jamesian pragmatism, Jungian depth psychology, Theologian and Philosopher Paul Tillich's conception of "ultimate concern," and fighter pilot John Boyd's OODA loop serves as philosophical, scientific and spiritual foundations for his teaching.

In 2015, he created *Story Grid Methodology* to begin teaching and further developing Simulation Synthesis Theory. Since then he has given lectures on the origin of story, the integration of storytelling and science, and the necessity of telling complex stories to thousands of students all over the world.

In addition to *The Story Grid* and *Mentoring the Machines,* he's authored, coauthored or ghost-written numerous bestselling nonfiction and fiction titles. His most recent lecture series, "Genre Blueprint" applies his Simulation Synthesis Theory to popular works such as *The Hobbit* by J.R.R. Tolkien and *The Matrix* by Lara and Lana Wachowski.

Printed in Great Britain
by Amazon